MEGA MACHINES

# Fire Engines

by Mari Schuh

Kaleidoscope
Minneapolis, MN

**Where the Quest for Discovery Begins**

*This edition first published in 2023 by Kaleidoscope Publishing, Inc.*

*For information regarding permission, write to*

*Kaleidoscope Publishing, Inc.*
*6012 Blue Circle Drive*
*Minnetonka, MN 55343*

*Library of Congress Control Number*
*2022937964*

*ISBN*
*978-1-64519-596-2 (library bound)*
*978-1-64519-666-2 (ebook)*

**Bigfoot Jr. lurks within one of the images in this book. It's up to you to find him!**

# Table of Contents

# Racing to the Fire

Gemma and her dad are traveling to the store. *New naw, new naw!* Gemma hears a loud siren and sees flashing lights.

"A **fire engine** is coming," Dad says. Dad drives the car out of the way, so the fire engine can pass.

The fire truck zooms by. It is fast!
It races to the fire.

**Firefighters** are ready to work. They fight house fires. They fight forest fires, too.

Fire engines carry water and hoses. Firefighters start to work. More help is on the way!

**Fire trucks** help fight fires, too. They have very long ladders. They can reach people in tall buildings.

Fire engine hoses are long and strong. They spray lots of water.

Fire engines carry **nozzles**. Nozzles connect to hoses. They help spray streams of water.

**Hoses are often different colors. This helps firefighters keep track of them.**

# Parts of a Fire Engine

window
lights
cab
door
wheel

Fire engines carry their own water. They have big water tanks. Strong pumps push water into the hoses.

Fire engines carry some **gear**. They carry supplies and tools. They have first aid kits, too.

**FUN FACT**

**Some fire engines carry air tanks, saws, and axes to help firefighters rescue people and stop fires.**

# Fighting Fires

Fire engines are tough trucks.
Strong tires go over rough land.
They get to the fire quickly.

Fire engines carry **foam**.
Firefighters spray foam on car fires.
They use it on chemical fires, too.

Firefighters work as a team. They work together to hold long hoses.

When firefighters need more water, they connect hoses to **hydrants**. Now they have more water to fight the fire.

**FUN FACT**
**Fire engines can also get water from pools and lakes.**

# Saving Lives

Gemma sees a fire engine on the road. “Where is it going?” she asks.

“It put the fire out and saved lives. Now it is going back to the fire station,” Dad says.

# Kinds of
# Fire Engines and Trucks

wildland

aerial

"The fire engine needs more supplies," Dad says.

"Then it will be ready to fight the next fire!" Gemma says.

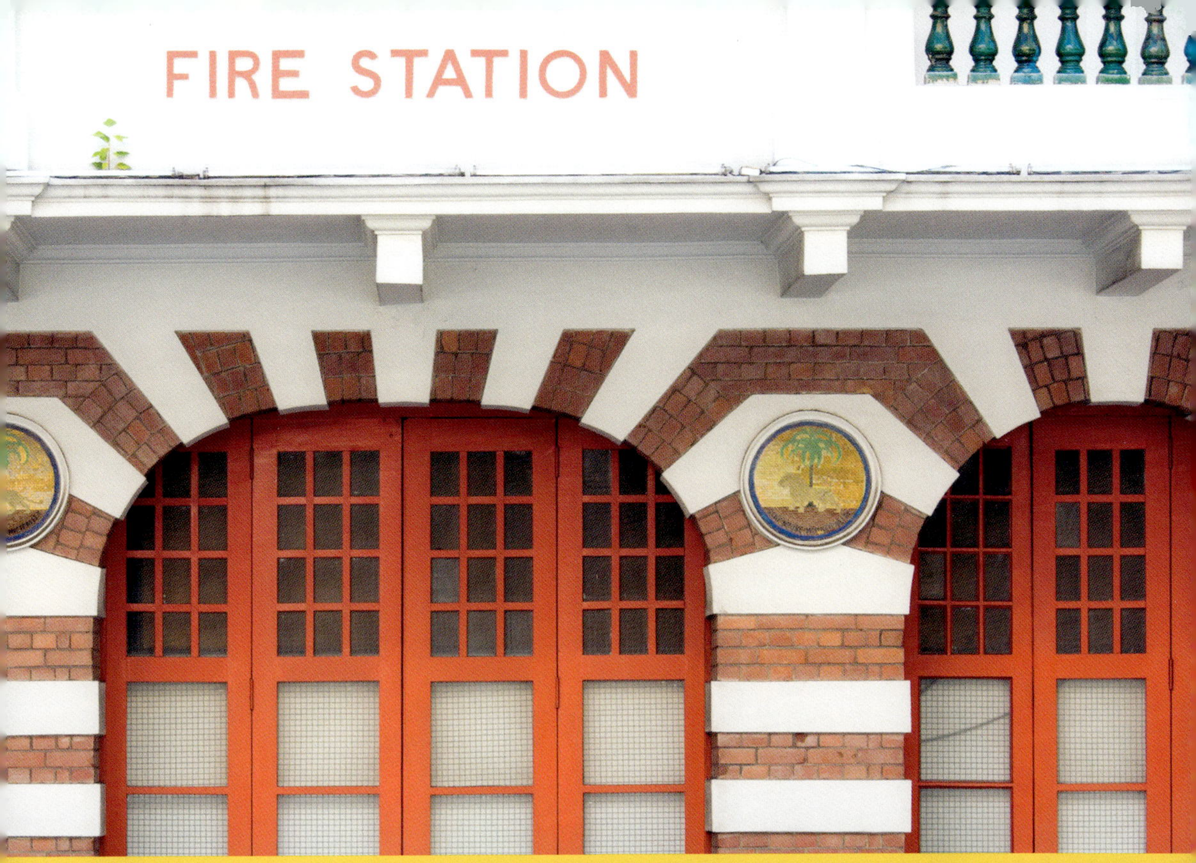
FIRE STATION

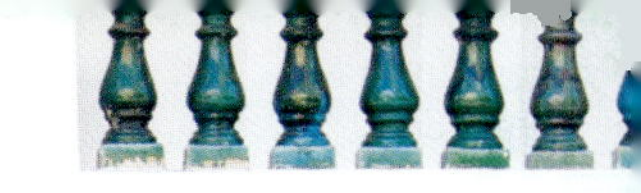

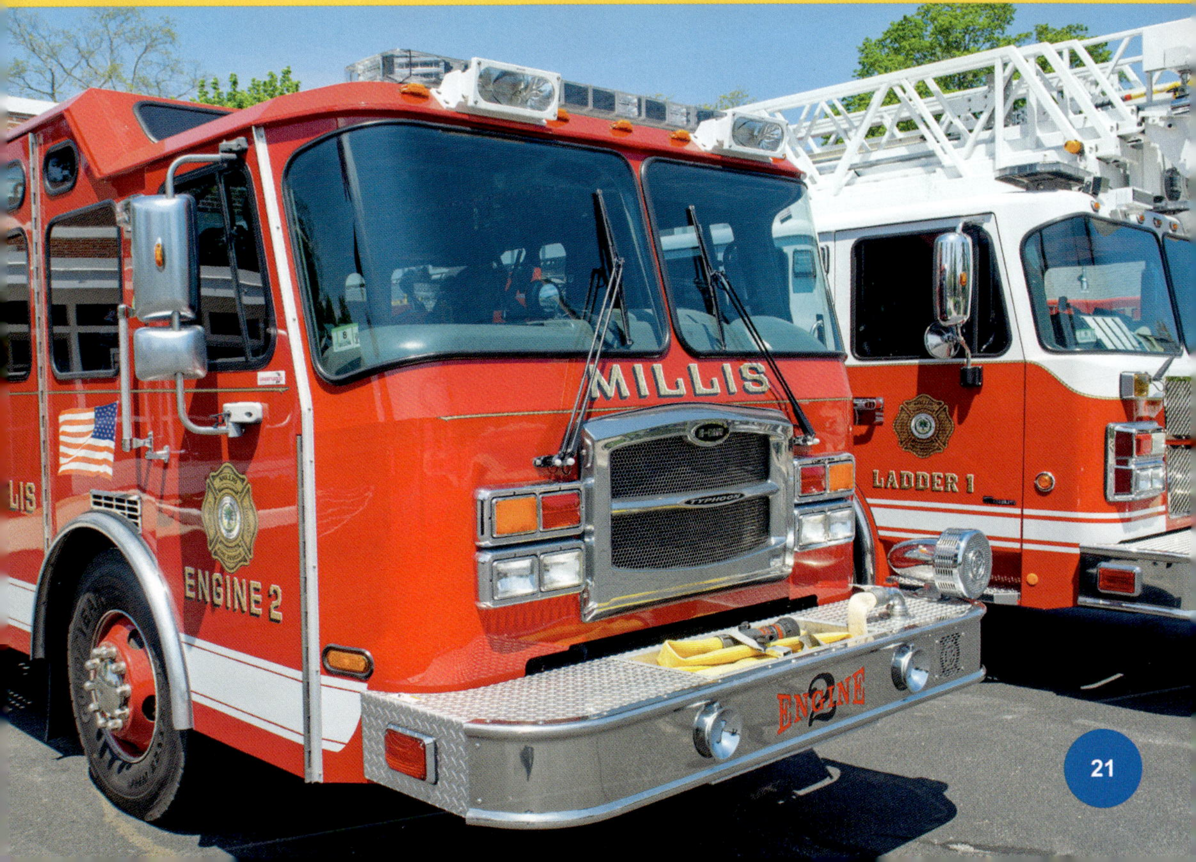
MILLIS
ENGINE 2
LADDER 1

# Photo Glossary

**firefighter:** A person who is trained to put out fires. Firefighters need to be strong and fit.

**fire engine:** A fire engine is a truck that carries water to fight fires. Pumpers are fire engines.

**fire truck:** A fire truck carries gear to fight fires and rescue people. Aerials are fire trucks.

**foam:** A mixture of small bubbles. Special foam is used to put out some fires.

**gear:** Equipment and clothing. Fire engines carry gear such as coats, boots, axes, and saws.

**hydrant:** A big pipe that sticks up out of the ground and is connected to a water supply. Some hydrant colors let firefighters know how fast water will flow from the hydrant.

**nozzle:** A part that directs the flow of water from the end of a hose. Nozzles can be set to speed up or slow down the flow of water.

# Read More

Besel, Jennifer M. *Fire Trucks*. Mankato, MN: Black Rabbit Books, 2023.

Dickmann, Nancy. *Fire Trucks*. North Mankato, MN: Pebble Books, 2022.

Harris, Bizzy. *Fire Trucks*. Minneapolis, MN: Jump!, 2022.

McDonald, Amy. *Fire Trucks*. Minneapolis, MN: Bellwether Media, 2021.

**FACTSURFER**

Factsurfer.com gives you a safe, fun way to find more information.

1. Go to www.factsurfer.com.
2. Enter "Fire Engines" into the search box and click
3. Select your book cover to see a list of related websites.

# About the Author

Mari Schuh's love of reading began with cereal boxes at the kitchen table. Today she is the author of hundreds of nonfiction books for beginning readers. Mari lives in the Midwest with her husband and their sassy house rabbit. Learn more about her at marischuh.com.

## INDEX



## PHOTO CREDITS

The images in this book are reproduced through Shutterstock: Rob Wilson 1; Tyler Olson 3, 22; Kekyalyaynen 4; Virrage Images 5; blurAZ 6; VAKS-Stock Agency 7, 9, 22; Kazela 8; sspopov 8; possohh 9; trekandshoot 10, 22, 23; robert_s 10; Matthew Strauss 12; Besim Bevrnja 13; Ralf Geithe 13; King Ropes Access 13; Moshe EINHORN 14; Kzenon 15; Prath 16; John Kasawa 16; Jerry Bergquist 17, 22; Barry Blackburn 18; 3D-Horse 19; Nerthuz 19; Flash Vector 19, 22; Firefighter Montreal 20, 22; Wangkun Jia 21; Cuson 21; S. Bonaime 22, SviatlanaLaza 22. Cover: Rob Wilson.